Umana Rafiq

Fourier Series Analysis And Applications

GRIN Verlag

Bibliografische Information der Deutschen Nationalbibliothek:

Die Deutsche Bibliothek verzeichnet diese Publikation in der Deutschen National-bibliografie; detaillierte bibliografische Daten sind im Internet über http://dnb.d-nb.de/ abrufbar.

Imprint:

Copyright © 2012 GRIN Verlag GmbH
Druck und Bindung: Books on Demand GmbH, Norderstedt Germany
ISBN: 978-3-656-73128-3

This book at GRIN:

http://www.grin.com/en/e-book/279395/fourier-series-analysis-and-applications

GRIN - Your knowledge has value

Der GRIN Verlag publiziert seit 1998 wissenschaftliche Arbeiten von Studenten, Hochschullehrern und anderen Akademikern als eBook und gedrucktes Buch. Die Verlagswebsite www.grin.com ist die ideale Plattform zur Veröffentlichung von Hausarbeiten, Abschlussarbeiten, wissenschaftlichen Aufsätzen, Dissertationen und Fachbüchern.

Visit us on the internet:

http://www.grin.com/

http://www.facebook.com/grincom

http://www.twitter.com/grin_com

Fourier Series Analysis And Applications

Umana Rafiq Ananna

Department of EEE, Ahsanullah University of Science and Technology, Dhaka, Bangladesh.

Abstract : **Fourier Series, Fourier Analysis, Euler's Formula for Coefficients, Periodic Functions, Trigonometric Series, Even Function, Odd Function, Properties of Functions, Fourier's Cosine And Sine Series, Half Range Fourier Sine and Cosine Series, Examples, Complex form, Riemann-Zeta Function, Mathematical analysis, Perseval's Formula, Piecewise smooth function, Bessel's inequality, Riemann lemma, Perseval's Theorem, Propositions and Remarks, Gibbs Phenomenon, Physical Applications, Heat distribution in a metal plate, Square wave, Sawtooth wave, Full an Half wave Rectifier, Advantages and Conclusion.**

1. Introduction

Fourier Series is the founding principle behind the field of Fourier Analysis, is an infinite expansion of a function in terms of sines and cosines. In Physics and engineering, expanding functions in terms of sines and cosines are useful because it allows one to more easily manipulate functions that are, for example, discontinuous or simply difficult to represent analytically. It decomposes periodic functions or periodic signals into the sum of a (possibly infinite) set of simple oscillating functions.

The Fourier series is named in honour of Jean-Baptiste Joseph Fourier (1768–1830), who made important contributions to the study of trigonometric series, after preliminary investigations by Leonhard Euler, Jean le Rond d'Alembert, and Daniel Bernoulli. Fourier introduced the series for the purpose of solving the heat equation in a metal plate.

These series are very powerful tools in connection with various problems involving Ordinary and Partial Differential equations. And have many more unique and undeniable real life applications which will be discussed as we proceed further.

2. Definitions

2.1 Periodic Functions

Let $f(x)$ be a real valued function said to be periodic if these exist in a non-zero number t, independent of x, such that the equation $f(x+t) = f(x)$ holds for all values of x. The least value of $t>0$ is called the least period or simply the period $f(x)$.

Example: Let $f(x)=\sin x$. Then $f(x)$ is a periodic function having period 2π.

Proof:
Here, $f(x) = \sin x$
$\Rightarrow f(x+2\pi) = \sin(x+2\pi) = \sin(2\pi+x) = \sin x$
$\Rightarrow f(x+4\pi) = \sin(x+4\pi) = \sin(4\pi+x) = \sin x$
. . . .
. . . .
$\Rightarrow f(x+2n\pi) = \sin(x+2n\pi) = \sin(2n\pi+x) = \sin x$ [n∈N]
$\therefore f(x) = f(x+2\pi) = f(x+4\pi) = \ldots \ldots \ldots = f(x+2n\pi) = \sin x$
Thus $f(x) = \sin x$ is a periodic function having period 2π.
Similarly we can show that $f(x) = \cos x$ and $f(x) = \tan x$ are periodic functions having period 2π and π respectively.

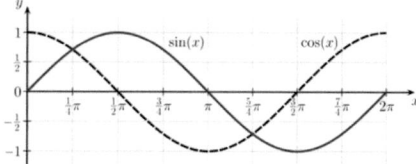

Figure1: The graph of sine and cosine function

2.2 Trigonometric Series
Any series of the term
$$\frac{a_o}{2} + \sum_{n=1}^{\infty}(a_n\cos nx + b_n\sin nx)$$
where a_o, a_n and b_n are coefficients and constants.

2.3 Fourier Series
The Trigonometric series
$$f(x) = a_o + a_1\cos x + a_2\cos 2x + \cdots + a_n\cos nx + \cdots + b_1\sin x$$
$$+ b_2\sin 2x + \cdots + b_n\sin nx + \cdots$$
$$f(x) = a_o + \sum_{n=1}^{\infty}(a_n\cos nx + b_n\sin nx)$$
is called a Fourier Series if the coefficients a_o, a_n and b_n are defined by the following formula which is called Euler's Formula for Coefficients ,
$$a_o = \frac{1}{2\pi}\int_{-\pi}^{\pi} f(x)$$
$$a_n = \frac{1}{\pi}\int_{-\pi}^{\pi} f(x)\cos nx dx$$
$$b_n = \frac{1}{\pi}\int_{-\pi}^{\pi} f(x)\sin nx dx$$
where, $f(x)$ is any single valued function defined on the integral $(-\pi,\pi)$. The Fourier series can also be written as
$$\frac{a_o}{2} + \sum_{n=1}^{\infty}(a_n\cos nx + b_n\sin nx)$$
where, $a_o = \frac{1}{2\pi}\int_{-\pi}^{\pi} f(x)$, $a_n = \frac{1}{\pi}\int_{-\pi}^{\pi} f(x)\cos nx dx$, $b_n = \frac{1}{\pi}\int_{-\pi}^{\pi} f(x)\sin nx dx$

2.4 Even Function

A function $f(x)$ is said to be even if $f(-x)=f(x)$ is symmetrical about y-axis.

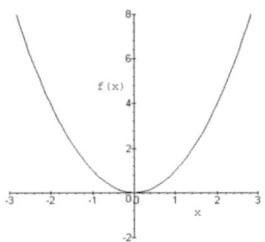

Example:
Let, $f(x)=x^2$
$\Rightarrow f(x)=(-x)^2= x^2$
$\therefore f(x)$ is an even function.

Properties Of Even Function: If $f(x)$ is Even then,

i. $\int_{-\pi}^{\pi} f(x)dx = 2\int_0^{\pi} f(x)dx$

ii. $a_0 = \frac{1}{\pi}\int_0^{\pi} f(x)dx$

iii. $a_n = \frac{2}{\pi}\int_0^{\pi} f(x)\cos nx\,dx$

iv. $b_n = 0$

2.5 Odd Function

A function $f(x)$ is said to be even if $f(-x)=f(x)$ is symmetrical about the origin.

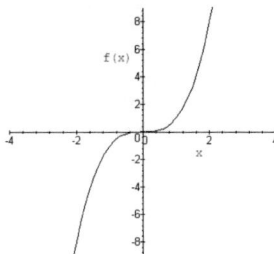

Let, $f(x)= x$
$\Rightarrow f(-x)=-f(x)=-x$ So, $f(x)$ is an odd function.

Properties Of Odd Function: If $f(x)$ is an odd function then,
i. $a_0 = 0$
ii. $a_n = 0$
iii. $b_n = \frac{2}{\pi}\int_0^{\pi} f(x)\sin nx\,dx$

2.6 Half Range Fourier's Cosine And Sine Series

When Fourier Series only has the Cosine terms or only the Sine terms we call such series as Half Range Fourier Cosine Series or Half Range Sine Series respectively. And such function must be defined in the integral $(0,\pi)$, which is half of the $(-\pi,)$ and the function is specified as odd or even so that it is clearly defined in the other half of the interval ,namely, $(-,0)$. In such a case we have,

$a_n = \frac{2}{\pi}\int_0^{\pi} f(x)\cos nx\,dx$ for half range cosine series and

$b_n = \frac{2}{\pi}\int_0^{\pi} f(x)\sin nx\,dx$ for half range sine series

2.7 Dirichlet's Condition For Fourier Series

If a function $f(x)$ for the interval $(-,)$
i. is single valued
ii. is bounded
iii. has at most a finite number of maxima and minima
iv. Has only a finite number of discontinuous
v. is $f(x)\infty+2 = f(x)$ for values of x outside $(-,)$ then,
$Sp(x) = \frac{a_0}{2} + \sum_{n=1}^{p}(a_n\cos nx + b_n\sin nx)$ converges to $f(x)$ as $p\rightarrow\infty$ at values of x for which $f(x)$ is continuous and to $\frac{1}{2}[f(x+0) + f(x-0)]$ at points of discontinuity.

2.8 Perseval's Formula

$\int_{-c}^{c}[f(x)]^2 dx = c\{\frac{1}{2}a_0^2 + \sum_{n=1}^{\infty}(a_n^2 + b_n^2)\}$

This is known as Perseval's formula.

1. If $0 < x < 2c$, then $\int_0^{2c}[f(x)]^2 dx = c[\frac{a_0}{2} + \sum_{n=1}^{\infty}(a_n^2 + b_n^2)]$

2. F $0 < x < c$ then (Half range cosine series) : $\int_0^{c}[f(x)]^2 dx = c[\frac{a_0}{2} + \sum_{n=1}^{\infty} a_n^2]$

3. F $0 < x < c$ then (Half range sine series) : $\int_0^{c}[f(x)]^2 dx = c[\frac{a_0}{2} + \sum_{n=1}^{\infty} b_n^2]$

4. R.M.S $= \left\{\frac{\int_a^b [f(x)]^2\,dx}{b-a}\right\}^{\frac{1}{2}}$

3. Work Examples

3.1 Fourier Series in Complex form
We Know,

$\cos(x) = \frac{1}{2}[e^{ix} + e^{-ix}]$

$\sin(x) = \frac{1}{2i}[e^{ix} - e^{-ix}]$

Now Fourier Series of a function $f(x)$ of period $2l$ is,

$$f(x) = \frac{a_0}{2} + \sum_{n=1}^{\infty}\left(a_n\cos\frac{n\pi x}{1} + b_n\sin\frac{n\pi x}{1}\right)$$

Now putting the value of cosx and sinx we get,

$$f(x) = \frac{a_0}{2} + \sum_{n=1}^{\infty}\left(a_n\frac{1}{2}[e^{ix} + e^{-ix}] + b_n\frac{1}{2i}[e^{ix} - e^{-ix}]\right)$$

Solving this equation we can get,

$$C_n = \frac{1}{2}(a_n - ib_n)$$

$$C_{-n} = \frac{1}{2}(a_n + ib_n)$$

where, $C_0 = \frac{a_0}{2} = \frac{1}{2}\frac{1}{l}\int_0^{2l} f(x)dx$

$C_n = \frac{1}{2}[\frac{1}{l}\int_0^{2l} f(x)\cos\frac{n\pi x}{1}dx - \frac{i}{i}\int_0^{2l} f(x)\sin\frac{n\pi x}{1}dx$

Hence,

$C_n = \frac{1}{2l}\int_0^{2l} f(x)e^{\frac{-in\pi x}{1}}dx$ and $C_{-n} = \frac{1}{2l}\int_0^{2l} f(x)e^{\frac{in\pi x}{1}}dx$

3.2 Riemann Zeta Function
Let $f(x) = x^2$ in the interval $[-\pi,\pi]$ .This is an even function so

we can have the Fourier Cosine series from this function.

Where, $a_o = \frac{1}{\pi} \int_0^\pi f(x)dx$

$a_n = \frac{2}{\pi} \int_0^\pi f(x)\cos nx dx$ and $b_n = 0$

Determining the coefficients and putting them in the Fourier Series of f(x) we get,

$$f(x) = x^2 = \frac{\pi^2}{3} + \sum_{n1}^{\infty} \frac{4}{n^2}(-1)^n \cos nx$$

or,

$$2 = \frac{2}{3} - \frac{4}{1^2} \qquad + \frac{4}{2^2} \qquad 2 \quad -\frac{4}{3^2} \qquad +$$

$$\frac{4}{4^2} \qquad + \cdots$$

Now putting $x = \pi$, and simplifying the equation we get,

$$\frac{2}{6} = \frac{1}{1^2} + \frac{1}{2^2} + \frac{1}{3^2} + \cdots \ldots \ldots = \sum_{=1}^{\infty} \frac{1}{2} = \xi(2)$$

Where $\xi(2)$ is called The Riemann-Zeta function.

4. Previous Results

The heat equation is a partial differential equation. Prior to Fourier's work, no solution to the heat equation was known in the general case, although particular solutions were known if the heat source behaved in a simple way, in particular, if the heat source was a sine or cosine wave. These simple solutions are now sometimes called eigensolutions. Fourier's idea was to model a complicated heat source as a superposition (or linear combination) of simple sine and cosine waves, and to write the solution as a superposition of the corresponding eigensolutions. This superposition or linear combination is called the Fourier series.

From a modern point of view, Fourier's results are somewhat informal, due to the lack of a precise notion of function and integral in the early nineteenth century. Later, Dirichlet and Riemann expressed Fourier's results with greater precision and formality.Some of the results obtained using Fourier Series are shown below.

4.1 Piecewise smooth function

A function f (x) is piecewise smooth on an interval if both f(x) and f(x)' are piecewise continuous on the interval.

Suppose that f(x) is piecewise smooth and periodic. Then the series with coefficients converges to
1. f(x) if x is a point of continuity.
2. if (x) is a point of discontinuity then,

$\frac{1}{2}(f(x+0) + f(x-0))$

This means that, at each x between -L and L, the Fourier series converges to the average of the left and the right limits of f(x) at x. If fis continuous at x, then the left and the right limits are both equal to f(x), and the Fourier series converges to f(x) itself. If f has a jump discontinuity at x then the Fourier series converges to the point midway in the gap at this point.
Remark: Let f(x) be a given piecewise continuous function. We say that f (x)is standardised if its values at points x_i of discontinuity are given by

$f(x_i) = \frac{1}{2}[f(x_i+) + f(x_i-)]$

Remark: The conditions imposed on f(x) are sufficient but not necessary, i.e if the conditions are satisfied the convergence is guaranteed. However, if they are not satisfied the series may or may not converge.

4.2 Bessel's inequality
Let f (x) be a function defined on $[-\pi,\pi]$ such that $f^2(x)$ has a finite integral on $[-\pi,\pi]$. If a_n and b_n are the Fourier coefficients of the function f (x), then we have

$$\pi(2a_0^2 + \sum_{n=0}^{\infty}(a_n^2 + b_n^2)) \leq \int_{-\pi}^{\pi} f^2(x)dx$$

In particular, the series $\sum_{n=0}^{\infty}(a_n^2 + b_n^2)$ is convergent.

Remark: The quantity $A_n = \sqrt{a_n^2 + b_n^2}$ is called the amplitude of the nth harmonic. The square of the amplitude has a useful interpretation. Indeed, borrowing terminology from the study of periodic waves, we define the energy E of a 2π-periodic function f (x) to be the number

$$E = \frac{1}{\pi} \int_{-\pi}^{\pi} f^2(x)dx$$

So Bessel's Inequality translates into:

$$(2a_0^2 + \sum_{n=0}^{\infty}(a_n^2 + b_n^2)) \leq E$$

4.3 Riemann lemma :
Let f be integrable and a_n and b_n be the Fourier coefficients of f. Then

$$\lim_{n \to \infty} a_n = \lim_{n \to \infty} b_n = 0$$

which means

$$\lim_{n \to \infty} \int_{-\pi}^{\pi} f(t) \cos nt dt = \lim_{n \to \infty} \int_{-\pi}^{\pi} f(t) \sin nt dt = 0$$

4.4 Perseval's Theorem :
If a function has a Fourier series given by

$$f(x) = \frac{1}{2} a_0 + \sum_{n=1}^{\infty} a_n \cos(nx) + \sum_{n=1}^{\infty} b_n \sin(nx), \qquad (1)$$

then Bessel's inequality becomes an equality known as Parseval's theorem. From (1),

$$[f(x)]^2 = \qquad (2)$$

$$\frac{1}{4} a_0^2 + a_0 \sum_{n=1}^{\infty} [a_n \cos(nx) + b_n \sin(nx)] + \sum_{n=1}^{\infty}\sum_{m=1}^{\infty} [a_n a_m \cos(nx)\cos(mx) + a_n b_m \cos$$

$$(nx)\sin(mx) + a_m b_n \sin(nx)\cos(mx) + b_n b_m \sin(nx)\sin(mx)].$$

Integrating

$$\int_{-\pi}^{\pi} [f(x)]^2 dx = \frac{1}{4} a_0^2 \int_{-\pi}^{\pi} dx + a_0 \int_{-\pi}^{\pi} \sum_{n=1}^{\infty} [a_n \cos(nx) + b_r \qquad (3)$$

$$\int_{-\pi}^{\pi} \sum_{n=1}^{\infty}\sum_{m=1}^{\infty} [a_n a_m \cos(nx)\cos(mx) + a_n b_m \cos(nx) s$$

$$a_m b_n \sin(nx)\cos(mx) + b_n b_m \sin(nx)\sin($$

$$= \frac{1}{4} a_0^2 (2\pi) + 0 + \sum_{n=1}^{\infty}\sum_{m=1}^{\infty} [a_n a_m \pi\delta_{nm} + 0 + 0 + b_n b_m \pi\delta_n$$

3

so

$$\frac{1}{\pi} \int_{-\pi}^{\pi} [f(x)]^2\, dx = \frac{1}{2} a_0^2 + \sum_{n=1}^{\infty} (a_n^2 + b_n^2). \tag{4}$$

For a generalized Fourier series of a complete orthogonal system $\{\phi_i\}_{i=1}^{\infty}$, an analogous relationship holds.

For a complex Fourier series,

$$\frac{1}{2\pi} \int_{-\pi}^{\pi} |f(x)|^2\, dx = \sum_{n=-\infty}^{\infty} |a_n|^2.$$

if a_n and b_n are the Fourier coefficients corresponding to f(x) and if f(x)satisfies the Dirichlet conditions.

4.5 The Gibbs Phenomenon
The partial sum of a Fourier series shows oscillations near a discontinuity point as shown from the previous example. These oscillations do not flatten out even when the total number of terms used is very large.
As an example, the real function f(x)=x has a value of 3.14 when x=3.14. On the other hand, the partial sum solution using 100 terms has a value of 0.318 when x=3.14. It is drastically different from the real value.
In general, these oscillations worsen when the number of terms used decrease. In the previous example, the value of the five terms solution is only 0.016 when x=3.14.
Therefore, extreme attention has to be paid when using the Fourier analysis on discontinuous function.
Near a point, where f has a jump discontinuity, the partial sums S_n of a Fourier series exhibit a substantial overshoot near these endpoints, and an increase in n will not diminish the amplitude of the overshoot, although with increasing n the the overshoot occurs over smaller and smaller interval. In this section we examine some detail in the behaviour of the partial sums S_n of

$$S(x) = \sum_{k=1}^{\infty} \frac{\sin kx}{k}$$

Let
$n \in \mathbb{N}$
and
$x_n = \frac{2\pi}{2n+1}$

$$\lim_{n \to \infty} \left(\sum_{k=1}^{n} \frac{\sin(kx_n)}{k} \right) = \int_0^{\pi} \frac{\sin \tau}{\tau}\, d\tau$$

and
$$\int_0^{\pi} \frac{\sin \tau}{\tau}\, d\tau = \frac{\pi}{2} \cdot 1.1789797\ldots$$

Since s(x) $\approx \frac{\pi}{2}$ for x near 0, we see that an "overshoot" by approximately 17.9% is maintained as n → ∞ (but over smaller and smaller intervals centred at x=0.

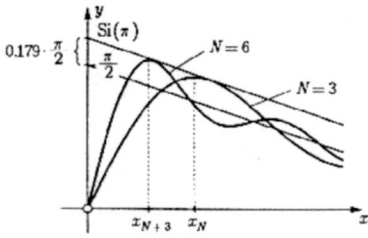

5. Main Results And Applications

Although the original motivation was to solve the heat equation, it later became obvious that the same techniques could be applied to a wide array of mathematical and physical problems, and especially those involving linear differential equations with constant coefficients, for which the eigensolutions are sinusoids. The Fourier series has many such applications in electrical engineering, vibration analysis, acoustics, optics, signal processing, image processing, quantum mechanics, econometrics,[5] thin-walled shell theory,[6] etc.

5.1 Using Fourier Expansion for square wave:
Using Fourier expansion we can analyze square waves. The analysis is important for electronic devices that are designed to handle sharply rising pulses.

Let us define a square wave by the function f(x) such that

$$f(x)= \begin{bmatrix} 0 & -\pi < x < 0 \\ h & 0 < x < \pi \end{bmatrix}$$

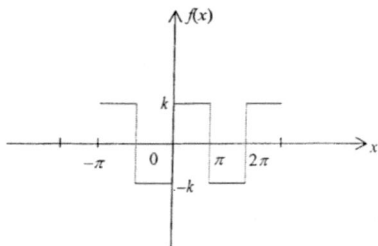

Which can be represented by Fourier Series as f(x) = a_0 + $\sum_{n=1}^{\infty}(a_n \cos nx + b_n \sin nx) \ldots (1)$
By determining the values of a_0, a_n and b_n then putting them in equation (1) we get,

$$f(x)= \frac{h}{2} + \frac{2h}{\pi} \left[\frac{\sin x}{1} + \frac{\sin 3x}{3} + \frac{\sin 5x}{5} + \cdots \right]$$

5.2 Fourier Expansion for Sawtooth wave:
We now use the formula above to give a Fourier series expansion of a very simple function. Consider a sawtooth wave

f(x) = x/π , for −π < x < π
f(x + 2πk) = f(x), for -∞ < x < ∞ and k∈Z

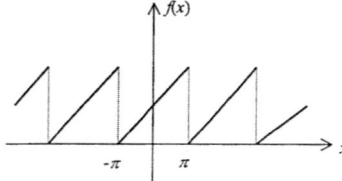

Then the Sawtooth wave can be represented by the Fourier Series by the following expression,

$$f(x) = \frac{a_o}{2} + \sum_{n=1}^{\infty} [a_n\cos(nx) + b_n\sin(nx)]$$

$$f(x) = 2\sum_{n=1}^{\infty}\frac{(-1)^{n+1}}{n}\sin(nx) \quad \text{for} \quad x\text{-}\pi \notin 2\pi Z$$

5.3 Full-wave Rectifier:

How well the output of a full-wave rectifier approaches to direct current can be determined by Fourier series analysis The rectifier which maybe thought of as having passed the positive peaks of an incoming sine wave and inverting the negative peaks, yields the function

$$f(t) = \begin{matrix} -\sin\omega t & -\pi < \omega t < 0 \\ \sin\omega t & 0 < \omega t < \pi \end{matrix}$$

And so the Function can be represented by Fourier series by the following expression,

$$f(t) = \frac{2}{\pi} + \frac{4}{\pi}[\frac{\cos 2\omega t}{3} + \frac{\cos 4\omega t}{15} + \ldots]$$

Where the original frequency w has been eliminated and the frequency of oscillation is 2ω. The high frequency components fall off as n^{-2}, showing that the full-wave rectifier does a fairly good job of approximating the direct current. Fourier series is also used to analyze the output of a Half-wave rectifier.

5.5 Heat distribution in a metal plate, using Fourier's method

One notices that the Fourier series expansion of our function in example 1 looks much less simple than the formula $f(x) = x/\pi$, and so it is not immediately apparent why one would need this Fourier series. While there are many applications, we cite Fourier's motivation of solving the heat equation. For example, consider a metal plate in the shape of a square whose side measures π meters, with coordinates $(x, y) \in [0, \pi] \times [0, \pi]$. If there is no heat source within the plate, and if three of the four sides are held at 0 degrees Celsius, while the fourth side, given by $y = \pi$, is maintained at the temperature gradient $T(x, \pi)$ $= x$ degrees Celsius, for x in $(0, \pi)$, then one can show that the stationary heat distribution (or the heat distribution after a long period of time has elapsed) is given by

$$T(x, y) = 2\sum_{n=1}^{\infty} \frac{(-1)^{n+1}}{n}\sin(nx)\frac{\sinh(ny)}{\sinh(n\pi)}.$$

Here, sinh is the hyperbolic sine function. This solution of the heat equation is obtained by multiplying each term of Eq.1 by $\sinh(ny)/\sinh(n\pi)$. While our example function $f(x)$ seems to have a needlessly complicated Fourier series, the heat distribution $T(x, y)$ is nontrivial. The function T cannot be written as a closed-form expression. This method of solving the heat problem was made possible by Fourier's work.

Conclusion

In this article, we have outlined the main features of the theory and application of one-variable Fourier series. Much additional information, however, can be found in the references. In particular, we did not have sufficient space to discuss the intricacies of multi-variable Fourier series. The Fourier Series is useful in many applications ranging from experimental instruments to rigorous mathematical analysis techniques. Thanks to modern developments in digital electronics, coupled with numerical algorithms such as the FFT, the Fourier Series has become one of the most widely used and useful mathematical tools available to any scientist.

References

1. Weisstein, Eric W. "Fourier Series." From MathWorld{A

Wolfram Web Resource
http://mathworld.wolfram.com/FourierSeries.html

2. Weisstein, Eric W. "Generalized Fourier Series." From MathWorld{A Wolfram Web Resource.
http://mathworld.wolfram.com/GeneralizedFourierSeries.html

3. Fourier Series and Their Applications.
Rui Niu, May 12, 2006
http://ocw.mit.edu/courses/mathematics/18-100c-analysis-i-spring-2006/projects/niu.pdf

4. Adanced Engineerg Mathematics, H.K. Das.
5. Wikipedia
http://en.wikipedia.org/

6. Fourier Series Analysis & Examples
http://www.sosmath.com

7. Encyclopedia of Physical Science and Technology, Fourier Series
James S. Walker, Department of Mathematics
University of Wisconsin–Eau Clair.

8. Mathematical Methods or Physiciss. Arkfen, Weber.

9. Boas, M. Mathematical Methods in the Physical Sciences, 2nd Edition.

10. Lectures On Fourier Series & Its Apliations, Professor Dr. Payer Ahmed.